CAREERS IN HVACR

HEATING, VENTILATION, AIR CONDITIONING, REFRIGERATION

HVACR TECHNICIANS KEEP PEOPLE WARM AND comfortable in the winter and pleasantly cool in the hottest of summer heat waves. From installing the machines and ducts that keep temperature-controlled air flowing through buildings, to maintaining vital air quality systems, these professionals do the work that lets the rest of us go through our days without threat of freezing or heat stroke.

HVACR is an acronym that stands for Heating, Ventilation, Air Conditioning, and Refrigeration. It is a trade that covers a variety of work. You may have had an HVACR technician come to your house to install or repair a furnace, air conditioner, boiler, or heat pump, but the field is much bigger than that. Many technicians choose to work in the commercial sector, which includes office buildings, shopping malls, factories, airports, and everything in between. They may install or service walk-in coolers in restaurants, hazardous gas and material ventilators in factories, massive cooling towers in high-rises, humidification (or dehumidification) units in hospitals, or air filtration systems on ships. Plus, there are literally dozens of specialized areas, from installing geothermal heating systems to energy auditing.

The income for doing this kind of work is attractive. With no experience at all, a beginner can start out at close to $30,000 a year. In less time than it takes to complete a college education, good HVACR professionals earn about $45,000 a year on average. Specialists earn even more – up to $80,000 a year, depending on their field and experience. Along with the income, a typical HVACR job provides benefits like insurance, holidays, and paid vacations. In many cases there are perks like company vehicles, cell phones, and laptops.

How hard is it to get started in this career? Anyone with mechanical aptitude can learn the trade on the job. Many do this through a formal apprenticeship program. Others simply start out as helpers, assisting an experienced technician who can teach them the skills of the trade. Still others go to vocational-technical schools for six to 12 months programs. Regardless of how much or little one prepares for a first HVACR job, the prospects are excellent. Labor experts predict as many as 100,000 new jobs will open up over the coming decade. These are jobs that cannot be automated or shipped offshore. Right now, there are not enough new technicians to fill empty positions. Jobs are everywhere for newcomers and experienced professionals alike.

HVACR is an excellent career choice for the mechanically inclined, especially those with an interest in electronics. It is a very large and diverse field where professionals are admired and respected.

WHAT YOU CAN DO NOW

HEATING, VENTILATION, AIR CONDITIONING and refrigeration is not a field that requires a great deal of formal education to get started. Much of an HVACR technician's training will take place through on-the-job training. However, the more you know ahead of time, the easier it will be to find a position as an inexperienced beginner. Plus, a little first-hand knowledge will go a long way towards increasing your earning potential right from the start.

High school students preparing for this career will need knowledge of math and physics. Concentrate on science and math, but do not neglect English and communications. Spanish language classes can be quite helpful as well. Any vocational and shop classes your school offers can be useful whether you become an HVACR technician or eventually pursue a different technical trade. Classes to consider include:

Wood shop

Metal shop

Drafting

Mechanical drawing

Sketching

Computer applications

Some knowledge of plumbing or electrical work and a basic understanding of electronics are also helpful. You can get this kind of experience by working part time after school helping a plumber or electrician.

A good way to explore whether this is the right career for you is to shadow an experienced technician on the job for a day or two. Be prepared to ask plenty of questions and gather advice on things you can do to prepare. Have your school counselor help you set this up.

Many HVACR technology programs offered by vocational-technical schools include paid apprenticeships. As an apprentice, you will get a realistic, first-hand look at the job while learning the skills of the trade. Apprenticeship requirements vary by area and local committee. Most programs require you to be at least 18, but some will take younger apprentices who have permission from their parents or high school principal.

If you cannot get into an apprenticeship and you are at least 16 years old, consider a part-time job after school or during the summer. Any job in construction will give you valuable work experience. There are opportunities for helpers and beginners at nonunion construction companies, especially during the busy summer months.

HISTORY OF THE CAREER

THE HISTORY OF THE MODERN HVACR SYSTEM is relatively short. There are those of us who can remember a time when an air-conditioning unit in the home was considered a luxury. Some homes today are still heated with wood burning stoves – not exactly a modern invention!

Looking back through ancient history will not produce any examples of HVACR systems with the kind of technology we enjoy today, but people have always needed some form of system to provide relief from uncomfortable temperatures. It started with the discovery of fire over 400,000 years ago, a major turning point in human history.

The key to obtaining warmth and protection on demand is the ability to control fire. There is scientific evidence that this ability was widespread approximately 125,000 years ago. Being in control of fire meant people could bring it inside their shelters and expand their activities into the dark and colder hours of the night. It would take almost 123,000 years – until modern times – to make much progress beyond having an indoor fireplace and man-powered fans.

The very first form of an HVACR system originated in ancient Rome over 2,000 years ago. Known as a *hypocaust*, it was a central heating system used for heating hot baths, houses, and public buildings. The system featured a furnace in a space beneath the floor, and flues inside the walls to distribute the heat. The design was revolutionary because hot air could heat the walls without polluting the interior of the room. Roman architect Vitruvius wrote about the hypocaust around the end of the first century BC, attributing the invention to Sergius Orata.

During the Renaissance, chimneys were widely used, allowing people to have comfortable, yet private, rooms. During this time, Leonardo Da Vinci devised a water driven fan to ventilate an entire suite of rooms, while the French started using ventilating machines in mines. In the late 1600s, Sir Christopher Wren made the very first gravity exhaust ventilation system for the House of the Parliament.

In 1714, German physicist Gabriel Daniel Fahrenheit invented the first mercury thermometer and devised the Fahrenheit temperature scale that is still used today. About 30 years later, Benjamin Franklin invented the first steam heating system. Known as the Franklin stove, his invention was used to protect people from the cold winters throughout the North American continent.

The 19th century saw a series of vital discoveries in the fledgling HVACR industry. Russian military engineer, Alexander Sablukov, invented the first ventilator using a centrifugal fan. It was quickly adapted for use in light industry, such as sugar production.

A stove with a furnace for heating air was introduced in England. The system was an arrangement of pipes that could heat up even big factories. Today the system is known as a direct fired heat exchanger.

In the latter part of the century, hot water heating systems were used for large commercial and public buildings. Houses were fitted with water spray systems for humidifying and cooling. Steam engines were used to drive air supply and exhaust air systems.

The 20th century saw major advances in the evolution of HVACR systems. The first fan coil dehumidifying system was made by a company called Buffalo Forge. The same company made the first spray type air-conditioning device and another system to remove dust particles from air streams.

In 1902, American engineer and employee of Buffalo Forge, Willis Carrier, invented air conditioning. Known as the father of modern air conditioners, Carrier built the first air conditioner to combat humidity inside a printing company – which it did when it was installed the very same year. Four years later he patented his invention under the name "Apparatus for Treating Air." The same year, Stuart Cramer coined the more user-friendly term, "air conditioning."

The first air conditioning for a home, a Carrier system, was installed in 1914. By the time World War II came to an end, everyone wanted A/C. By 1953, room air conditioner sales exceeded one million units, and manufacturers could not keep up with the demand.

There was, however, a downside to widespread use of air conditioning – the over usage of refrigerants endangered the environment. In 1987, the United Nations Montreal Protocol for protection of the earth's ozone layer was signed. The Protocol established unprecedented international cooperation for the phasing out of chlorofluorocarbon (CFC) refrigerants. CFC production in the US officially stopped on December 31, 1995. A short time later, "Research for the 21st Century" began. The object of this long-term, multi-million dollar program is to research air conditioning and refrigeration equipment to find ways to decrease energy usage while improving indoor air quality.

Today's HVACR systems have sensors that can communicate with the thermostat, meters, and apps that are accessible on the web or smartphones. These smart systems can lower energy usage a whopping 20 percent immediately upon installation. In addition, their ability to communicate in real time with homeowners and building managers can push that efficiency up to 30 or even 40 percent. The systems also send out alerts when a part breaks or refrigerant leaks.

Even the most intelligent machines need trained technicians to manage them. The future of smart HVACR systems relies on qualified professionals to install and service them. Regulations across the country require smarter cooling and heating in new buildings, and more energy efficiency. HVACR technicians will be designing and installing smart systems in new homes, as well as retrofitting the millions of homes and offices built over the past decades which need upgrading.

WHERE YOU WILL WORK

HVACR TECHNICIANS CAN BE FOUND WORKING wherever there is equipment used to control the climate – in residential, commercial, and public facilities of all kinds. More than half are employed by construction, plumbing, heating, and air-conditioning contractors to perform new installations. Some work for fuel oil and gas dealers, refrigeration and air-conditioning service and repair shops, and stores that sell heating and air-conditioning systems. Local, state, and federal governments hire HVACR technicians to install and repair systems in schools, airports, and other government buildings. Hospitals, office buildings, and other organizations that operate large air-conditioning, refrigeration, or heating systems usually employ HVACR technicians on staff. About 10 percent are self-employed. The rest are employed by an assortment of surveying companies, energy management firms, manufacturers, and engineering firms.

HVACR contractors and technicians who want to enjoy plentiful job opportunities, high salaries, available training, and an abundance of service calls should head for California or Ohio. California has more certified technicians than any state, and the highest employment (16,000). Ohio has fewer sunny days than California, but it boasts the highest number of accredited HVACR schools in the country, plus a high number of equipment wholesalers that employ technicians.

These two states top the list of the 10 best states to work in HVACR. The other eight states are:

Florida

Texas

Illinois

New York

Pennsylvania

New Jersey

North Carolina

Georgia

In addition, any states running up and down the East Coast generally have a higher number of HVACR technician jobs.

The work itself is done both inside and outside, in warm weather and cold weather. Inside, the work can be uncomfortable because of cramped or awkward workspaces or the need to perform maintenance on parts in high locations. The building may also be uncomfortable because the air-conditioning or heating system is broken, or has not even been installed yet. Outside there is plenty of room to work, but the weather does not always cooperate. The work must generally go on despite heat, cold, wind, or rain.

This is a very stable industry, work goes on throughout the year, especially since a growing number of manufacturers, and contractors now provide or require year-round service contracts. The majority of HVACR technicians, those who primarily do installations, work a standard 40-hour workweek, with only occasional evening or weekend shifts during peak seasons. Maintenance workers, however, often work evening and weekend shifts and may also be on call during off hours.

THE WORK YOU WILL DO

IMAGINE SPENDING THE SUMMER in Phoenix with no air conditioning, or the winter in Fargo with no furnace. Would it even be possible to live without succumbing to heat stroke or frostbite? Maybe, but we have all come to expect more than just survival. We want to live in comfort no matter what Mother Nature throws at us. We take for granted the heating and cooling systems that control the temperature, humidity, and air quality in our homes and businesses. In additional to issues of comfort, they are also vital for transporting and storing food, medicine, and other perishable items.

Heating and cooling devices are known as HVACR systems. HVACR is the acronym for heating, ventilation, air conditioning, and refrigeration. The workers who install, maintain, and repair the systems are generally referred to as HVACR technicians.

HVACR systems consist of many different components. Electrical, mechanical, and electronic parts may include thermostats, fans, motors, compressors, pumps, and switches. There are non-mechanical parts, too, like pipes and ducts. For example, a central air-conditioning system will use a compressor to cool the air before distributing the air. Then a fan blows the cooled air throughout the building via a network of metal or fiberglass ducts. If any part of the system malfunctions, an HVACR technician must be able to identify the problem. This is typically done by adjusting system controls and running performance tests using special tools and test equipment. Once the problem is diagnosed, the technician is able to carry out repairs on any part of the system that is not functioning correctly.

HVACR technicians will often specialize in heating, cooling or refrigeration, though technicians are usually trained for all job functions. Each type of system poses different concerns regarding fuel types and maintenance issues. Heating technicians have to make sure that systems are not leaking hazardous and flammable fuel, and that dust does not build up that could potentially catch fire if heated excessively.

Technicians working with cooling and refrigeration devices must handle hazardous coolants and be careful to dispose of these fluids in a way that will not harm the environment. Some HVACR technicians work solely with one type of device such as gas burners, water-based heating systems, solar panels, or commercial refrigeration.

All HVACR technicians are trained to do the following:

Use blueprints and design specifications to install HVACR systems

Connect systems to fuel and water supply lines, air ducts, and other components

Install electrical wiring and controls and test for proper

operation

Inspect HVACR systems during routine scheduled maintenance

Test individual components and diagnose problems in malfunctioning systems

Repair or replace worn or defective parts

Determine an HVACR system's energy use and make recommendations to improve efficiency

This work requires a large variety of tools. Some are common hand tools, such as screwdrivers, pipe cutters, hammers, metal snips, and wrenches. These tools can be used in both installations and maintenance. There are also many specialized tools such as carbon monoxide testers, voltmeters, combustion analyzers, measurement gauges, manometers, and acetylene torches. These tools are more likely to be used during maintenance to test air flow, refrigerant pressure, electrical circuits, burners, and other components.

Although HVACR technicians are trained to both install and maintain systems, many focus on either installation or maintenance. This is often the case for those who work for large contracting companies or directly for a manufacturer or wholesaler. Those working for smaller operations tend to do both installation and servicing, and work with all kinds of systems – heating, cooling, and refrigeration.

Installations

HVACR technicians who specialize in the installation of cooling and heating systems typically work on residential or commercial buildings, but not both. Those doing residential installations are usually employed by building contractors to install systems in new homes while they are being built. Some work for manufacturers or their distributors, removing old systems to upgrade with new, more efficient systems.

On the commercial side, technicians handle large installations in office buildings, retail stores, public facilities, factories, and other industrial buildings. Most of these are covered by union contracts, which changes the type of work the HVACR technician does. For example, sheet metal workers might do the ductwork while electricians do any electrical work. Pipefitters or steamfitters may install piping, while plumbers install the condensers. The HVACR technician pulls all these components together to complete the systems.

HVACR technicians who install heating systems are usually referred to as furnace installers. These technicians follow blueprints or other specifications to install oil, gas, electric, solid-fuel, and multiple-fuel heating systems. To ensure the proper functioning of the system, they often use combustion test equipment such as carbon dioxide and oxygen testers.

HVACR technicians who install central air-conditioning systems are known as refrigeration technicians. These technicians also read blueprints, but more often they follow design specifications and manufacturers' instructions to place refrigeration equipment. They start by bringing in and positioning all the components, including pumps, motors, compressors, condensing units, and piping. These components are then connected to air ducts, refrigerant lines, vents, water supply lines, and finally, the electrical power source. Once everything is assembled, they charge the system with refrigerant, check it for proper operation, and program the control systems.

Maintenance

Historically, contracting businesses, manufacturers, and wholesalers have employed HVACR technicians to do maintenance. That is because maintenance contracts were needed by commercial and industrial establishments to ensure their operations were not affected by system failures. Today, however, service contracts for smaller companies and residential customers are gaining popularity. In fact, it is often the installation technicians who sell the regular service contracts to their customers. This practice is advantageous for both installers and maintenance technicians. It provides extra income for the installer who is paid a commission for each contract sale, and it also provides off-season work for the maintenance technician who typically sees natural fluctuations throughout the year.

A typical service contract usually includes the cleaning of ducts, replacing filters, and checking refrigerant levels. After a furnace has been installed, the heating equipment needs routine care to continue operating efficiently. During the fall and winter, when the system is being used the most, the technician will check and adjust the burners and blowers to make sure they are functioning properly. If the system is not operating properly, the thermostat, burner nozzles, controls, or other parts are checked to identify and then correct the problem. During the summer, when the heating system is not being used at all, the technician will do routine maintenance, such as replacing filters, ducts, and other parts of the system that may have collected dust and impurities while operating over the winter.

Maintenance for cooling systems is similar except that it is scheduled for opposite seasons from heating systems. Repair calls are done during the warm weather months, while routine maintenance, such as overhauling compressors, is done in the winter.

When working on air conditioning and refrigeration systems, technicians must follow government regulations regarding the conservation, recovery, and recycling of refrigerants. This includes the proper handling and disposal of fluids and pressurized gases. Technicians conserve refrigerants by making sure that there are no leaks in the system. They recover it by venting the refrigerant into proper cylinders. They recycle it for reuse.

New technology, in the form of smart phones, allows technicians to tap into the Internet to diagnose problems. Computer hardware and software have been developed that allow heating, venting, and refrigeration units to automatically alert the company responsible for maintenance when problems arise. The central office can then notify the technician in the field via cell phone. The technician can then access the Internet to "talk" with the unit needing maintenance. This technology not only saves time and makes the job easier for the technician; it also represents substantial savings for the customer.

HVACR TECHNICIANS TELL THEIR OWN STORIES

I Service HVACR Systems in Homes

"When I was in high school my father gave me some great advice. He said when choosing a career, make sure it is one that serves a need. I knew I was mechanically inclined so I naturally looked at the various trades. Keeping my father's words in mind, I chose HVACR because the need for heating and air conditioning will never go away. People can live without a lot of luxuries, but when the A/C goes down in the summer or the furnace quits in January, people want it back immediately! That translates to job security for me, not to mention the satisfaction of being able to satisfy people's need for comfort.

The HVACR field is divided into two main categories, commercial and residential. Both of these categories are then divided into service and installation. Breaking into the commercial side typically requires some higher education or an apprenticeship program. The residential side is somewhat easier to get into. I chose the residential side because I wanted to get started right away, earning my way up through the ranks. I wasn't big on sitting in a classroom any longer than necessary. I started out as an installer's helper, but soon learned that the workflow can be very dependent on the amount of new construction taking place at the time. The service side, on the other hand, provides a steadier workflow. It only took me a month to find a job on the service side.

A good HVACR technician is a 'jack of all trades.' Over the years, I gained a lot of knowledge of other areas such as electrical, plumbing, and framing, as well as the newer green industry systems that use solar and geothermal energy. I've learned to use hundreds of tools from the basic screwdriver to a refrigeration recovery unit.

One of the best parts of my job is the endless variety. I get bored easily and could never be happy going to the same job site day after day. As an HVACR service tech, there is no boredom or monotony. Every day is a surprise, with new places to work, new people to meet, and new issues to resolve. There is always something new to learn, too, whether it's new building codes, OSHA regulations, or electronic control systems.

HVACR is a great career choice for anyone who wants

to feel a sense of accomplishment and success at the end of the day. Yes, there is some hard work and danger involved, but by using the proper tools and following safety procedures, this can be a long and rewarding career."

I Am an HVACR Engineer

"HVACR engineering is more hands-on than most other kinds of engineering. I rarely sit at a desk and I don't wear a tie. Instead, I throw on a hard hat and go into a job site, which is usually an older commercial building of some sort.

I am known as 'Mr. Fixit' because I'm a troubleshooter. Engineering requires a mechanical aptitude, but as a troubleshooter the two most important skills are problem solving and communication. In fact, communications skills are valued above all in my field. Troubleshooting is a process of elimination, ruling out certain factors and zeroing in on the source of the problem as quickly as possible. As I go along, I need to speak and write clearly so my co-workers, clients, and product vendors understand what's going on and what needs to happen.

Engineering is a good choice for someone who enjoys working with the latest gadgets. Most of the tools I use every day are software based. The most common task is running simulations of energy performance and load calculations. I also use software to predict the temperature and pressure inside pipes and ducts. I still use handheld measuring devices and data loggers, but the computer is my go-to tool.

A lot of HVACR pros like to install clean, new systems with all the latest features. Most don't want to deal with problems and wouldn't know where to start anyway. I love problem solving. A fun day for me is digging into a factory that was built in the '80s, getting to know its history, and figuring out where and when things started to go wrong. There are many reasons an HVACR system can stop working correctly. It's very satisfying to keep eliminating possibilities and finally come up with the solution.

What I enjoy most is the reaction I get from clients, especially when it's been a chronic problem that has become really bad. The building operations people are really grateful when I come and finally solve the problem. I feel a great sense of pride when I fix something that's been broken for a long time and no one else could figure it out. This is a field where you can have an impact every day. As an HVACR engineer, I know I can make a difference."

I Am an Energy Auditor

"I help people conserve and save energy, and show people how their energy bills will go down significantly if they upgrade or replace their old HVACR systems. I started out as a technician, but as more people became concerned with energy conservation, I found myself spending a lot of time educating people on what they could do to save money while helping the environment. I enjoy working with people, so I spent a few weeks getting the necessary training and obtained my certification as an energy auditor.

I like variety in my workday and this job gives me that. My schedule is an equal mix of working in the office and out in the field. I spend half the day making calls, following up on work orders, and filling out paperwork. The other half day I'm out performing audits, usually three on any given day.

Working with people is the fun part of the job. Homeowners come from all different backgrounds so I never know who I'm going to meet. One thing is certain, they all like talking about their home. They are happy to show me the unique things about their homes and have me check out all the nooks and crannies. It's like being an investigator, looking for the clues, then solving the case.

I usually work solo, but about once a week I will have someone shadowing me, helping out while learning what this field is all about. I encourage anyone who is interested in the field to check it out. The job prospects are good and just keep getting better. A lot of companies will even pay for training.

The pay is good and the work is essential. What I do affects how people live and work. I don't think you can ask much more from any career."

PERSONAL QUALITIES

TO WORK IN HVACR, YOU MUST have a solid grasp of the basics. It is also an ever-evolving trade with new materials, techniques, and innovations coming along every day. To be successful, a technician must embrace change and be willing to continually learn new things long after completing the initial training.

Although the work is technical in nature, people skills are important. HVACR technicians often deal directly with the public, especially those who work in customers' homes or business offices. A technician should always be friendly and courteous. That may sound easy, but customers can get quite cranky when temperatures are outside the normal comfort zone. Dealing with an unhappy customer whose air conditioning is not working on a hot July day takes patience, tact and a bit of humor.

Physical strength is needed for this work. HVACR technicians sometimes have to lift and support heavy equipment and components, often without help. They also have to work in cramped spaces like attics or squeeze through crawlways under buildings. Staying in good physical condition can mean fewer aches, pains, and strains. You should be prepared for less than ideal working conditions, too. Work is sometimes done outside, where you will be exposed to all sorts of weather. If you are afraid of heights or confined spaces, you could have a problem.

Troubleshooting skills are essential. Heating, air conditioning, and refrigeration systems are comprised of many intricate parts. If any one of those components malfunctions, the entire system can go down. All HVACR technicians are trained to conduct various tests to identify problems. It can take a while to run through all of them. A good technician pays attention to tiny clues that can lead the way to a correct diagnosis quickly.

Being detail oriented is a big plus. For starters, paperwork is part of the job. All work performed must be carefully recorded to include information like how long it took to do the work, what actions were taken, what specific parts were used, and when the equipment should be serviced next. Leaving out a single detail could mean trouble down the road.

You must also be able to follow instructions exactly, whether they come from your boss or technical drawings and building plans. You will need to follow building codes and standards and keep up on the latest changes in legislation. Getting sloppy could lead to costly mistakes.

HVACR technicians install and work on complicated climate-control systems. You do not have to be a computer genius, but should at least have an aptitude for math, science, and engineering. Electronics are becoming more common in the HVACR field, so the sooner you get comfortable with the subject the better.

ATTRACTIVE ASPECTS

THE HVACR INDUSTRY IS A VERY LARGE, global business that promises job security, excellent earnings potential, and opportunities for advancement. It is an ideal field for those who like to work with their hands, but it is an exciting industry that extends well beyond wrenches and gauges.

HVACR professionals cite job security as the number one reason they chose this career path. Indeed, the job growth in the HVACR field is exceptional. Experts project the field will see a 34 percent increase in demand over the coming decade. That translates to a massive number of new jobs, as many as 100,000 in just 10 years! These are jobs that cannot be exported or outsourced, and they are recession proof. Jobs are everywhere with opportunities for both qualified technicians and hardworking newcomers nationwide.

HVACR technicians enjoy good earnings. The average worker's pay is inching close to $50,000 a year and some industry experts project a pay increase of up to 40 percent in the next few years. It is a simple matter of supply and demand. There are more jobs opening up than employers can quickly fill. The pressure to fill those jobs will push up wages. Plus, most jobs come with good benefits and the opportunity to earn even more through overtime pay and sales bonuses.

This is a diverse career with many different positions for both the trained and untrained worker. From a job on the line to a desk in the research and development lab, it takes many positions to develop, engineer, build, produce, package, market, install, maintain, and sell an HVACR product. At every one of those steps there is a job that needs to be completed. You can change course at any time and use the education and experience already gained to move through different types of positions and responsibilities.

There are many opportunities for advancement. The HVACR industry is arguably one of the best trade fields to work in if you are ambitious. The key to success is exploring its many facets to find the best career path for advancement. Someone may start out as a technician or a dispatcher, but where they advance to is up to them. Typically, HVACR technicians can advance from service technicians to installers, operation managers, and distribution managers. Those with a head for business and good management skills can even start their own companies.

The HVACR industry is perfect for someone who enjoys the hands-on work of a trade, but wants to be involved in science. This is a technology-driven field that is on the cutting edge of emerging green technologies, for example, such as geothermal and solar. There is a growing need for technicians with strong electronics, controls, and networking skills. In fact, many HVACR technicians today use electronic tools and computers more than they do wrenches.

UNATTRACTIVE ASPECTS

HVACR CAN TAKE ITS TOLL ON A BODY. Performing any kind of physical labor is draining, both physically and mentally. It can lead to high stress levels and burnout over time. HVACR work can be dirty, sweaty, freezing, and wet. Working in attics is probably the worst part of this business. Squeezing into tight spots where you will constantly bang your head is only part of the problem. In the summer, when a high percentage of service calls occur, the attic is roasting. When working in tight spaces, inhalation of refrigerants is also a risk.

The work can be hazardous. In fact, HVACR technicians have one of the highest rates of injuries and illnesses of all occupations. Potential hazards include electrical shock, burns, muscle strains, and other injuries from handling heavy equipment.

One of the main culprits is refrigerants. Technicians have

to wear appropriate safety equipment because contact with refrigerants can cause skin damage, frostbite, or even blindness. The situation does not seem to be getting any better either. Just in the last few years, several refrigerants have been introduced that are highly flammable, requiring additional care.

Training never stops. If you like to feel there is nothing new to learn, this is not the career for you. HVACR is constantly evolving. Technicians must keep up with the latest products, tools, techniques, and technologies.

This is a stable industry loaded with opportunities for both trained and untrained workers. However, beginners looking for their first jobs may find it difficult to get their first job, especially if they are looking in the off-season months. Most apprentices are hired to help experienced technicians that are busy during months of extreme temperatures, both hot and cold. It is advisable to time your career start accordingly.

EDUCATION AND TRAINING

TRADITIONALLY, HVACR TECHNICIANS have learned the trade through on-the-job training. While this is still possible, it is becoming much less common. Those who do, typically start out as helpers, assisting experienced technicians. Appropriate tasks for beginners are very basic, such as insulating refrigerant lines, carrying supplies, or cleaning furnaces. As they gain experience, they move on to more difficult tasks, such as cutting and soldering pipes or checking electrical and electronic circuits.

Because HVACR systems are becoming increasingly

complex, the on-the-job training route may not be the best choice for a new careerist. The training is limited to the employer's knowledge and skills, which may be out of date. Moving on to another employer is not necessarily the answer either. A growing number of employers will only hire those with formal training from a technical school or apprenticeship program.

Finding an HVACR training program is easy. There are many vocational technical schools and community colleges – plus the Armed Forces – to choose from. Completing a program can take as little as six months, and no more than two years. Students learn the basics of installation, maintenance, and repair of heating, air-conditioning, and refrigeration systems. Instruction also covers theory, design, and equipment construction, as well as microelectronics, which are becoming increasingly important. Depending on the program and where it is offered, a graduate will receive either a certificate or an associate degree.

Apprenticeships

Most technicians today obtain their training through an apprenticeship. Apprenticeship programs are typically run by joint committees representing local chapters of national unions, such as the Air Conditioning Contractors of America and the Mechanical Contractors Association of America. To get accepted into an apprenticeship program, an applicant must meet these requirements:

Be at least 18 years old

Have a high school diploma or equivalent (GED)

Pass a basic math test

Pass substance abuse screening

Have a valid driver's license

Have good reading skills

Apprenticeship programs usually last three to five years. Each year, the apprentice must complete 2,000 hours of on-the-job training plus a minimum of 144 hours of classroom instruction. First year classes cover basic subjects such as the use and care of tools, safety practices, and blueprint reading. As the program progresses, apprentices also learn about the numerous systems and their components by studying:

Automated HVACR controls

Motors and motor controls

Steam and hot water systems

Duct layout and fabrication

System testing and troubleshooting

Cold-water air conditioners and domestic appliances

Electrical circuitry

Installation procedures

Air quality problem solving

Industrial and commercial refrigeration systems

Certifications and Licenses

Whether HVACR technicians learn the trade on the job or through some type of formal training program or apprenticeship, they can take several different tests to testify to their skills. Certifications can be helpful because they demonstrate to prospective employers that a technician has specific competencies. Many industry organizations offer certifying exams, which are usually conducted through vocational technical schools. Several organizations have begun to offer basic self-study, classroom, and Internet courses for those with limited experience.

There are different tests for different levels of experience. Technicians with relevant coursework and fewer than 2 years of experience are eligible to take the entry-level certification exams, such as the Industry Competency Exam or the Secondary Employment Ready Exam. These exams test basic competency in residential heating and cooling, light commercial heating and cooling, and commercial refrigeration.

Those with at least one year of installation experience and two years of maintenance and repair experience can take a number of specialized exams. Usually, a technician takes one of these exams to prove competency with specific kinds of equipment such as compressed-refrigerant cooling systems or oil-burning furnaces.

The industry recently adopted a single standard for certification of experienced technicians – the Air-Conditioning Excellence program, which is offered through North American Technician Excellence (NATE).

Some states and localities require HVACR technicians to be licensed. Although specific licensing requirements vary, all candidates must pass an exam.

In addition, the US Environmental Protection Agency (EPA) requires any technicians who purchase or work with refrigerants to be certified in their proper handling. There are numerous vocational technical schools, unions, and industry associations that provide training programs specifically designed to prepare a technician for the EPA exam. Technicians seeking certification must pass a written exam specific to the type of work they will be doing. The three possible certifications are: Type I – servicing small appliances, Type II – high-pressure refrigerants, and Type III – low pressure refrigerants.

EARNINGS

HVACR TECHNICIANS HAVE VERY GOOD EARNINGS potential. The average HVACR technician earns over $45,000 a year. That is just the average, which includes everyone from beginners with absolutely no training or experience, to highly experienced professionals. Those who excel and specialize in a sought after area are in the top 10 percent of earners with annual salaries of around $75,000, and even that is just an average. HVACR technicians who work for large companies often make more than $85,000 a year. Some even earn over six figures, usually by earning an HVACR related degree or starting their own HVACR company.

The average starting salary for a new beginner or apprentice is around $26,000. As they gain experience and improve their skills, apprentices receive periodic raises until they reach the salary level of experienced workers.

In addition to a good salary, a typical HVACR job will come with a variety of benefits. Health insurance, paid vacations and holidays, and pension plans are fairly standard. There may be other perks as well. Some companies pay for work-related training and provide uniforms, company vans, and tools. Some will go a step further and provide cell phones and laptops.

The best bonus money goes to those with a knack for sales. It is very common for companies to offer commission opportunities to its employees because sales are where most of the company revenue comes from. Getting a certain percentage of a sale of a new system can mean as much as $2,500 – a major plus.

This is an industry where it pays to join a union. More than 20 percent of HVACR technicians are members of a union. Although there are plenty of smaller unions around, the two unions with the biggest memberships are the Sheet Metal Workers' International Association and the United Association of Journeymen and Apprentices of the Plumbing and Pipefitting Industry of the United States and Canada.

Union members are protected under federal legislation and often receive higher wages as a result. Plus, unions negotiate the terms and conditions of employment contracts for their members, and often are able to get benefits packages that are significantly better than those offered to non-union workers.

OPPORTUNITIES

JOB GROWTH WITHIN THE HVACR INDUSTRY is one of the best reasons to consider this career. The field is expected to produce a very healthy 20 percent increase in employment over the coming decade. That is more than the average for almost any other occupation. HVACR is an industry that thrives even in times of a poor economy. Technicians enjoy job security and little fear of layoffs. HVACR work cannot be exported abroad or automated to robots, so there will always be a need for qualified professionals.

There are many factors behind this favorable outlook, starting with basic attrition. There is a high percentage of technicians today who are reaching retirement age, and there are not enough new technicians entering the trade to replace them. There is the resurrection of the construction industry that took a big hit during the recent recession. The majority of HVACR technicians are employed by commercial and residential building contractors. As the construction industry continues to recover, it drives employment growth for all trades, including HVACR.

There is also the fact that climate-control systems generally need replacement after 10 to 15 years. This means that homes and commercial buildings that were constructed between 2000 and 2005 need replacement systems right now. Further spurring the demand for technicians is the growing number of businesses that use refrigerated equipment, such as supermarkets, restaurants, and convenience stores.

Employers are especially interested in hiring technicians who are familiar with electronics and computers. Many technicians today spend more time using electronic tools and industry-specific computer software than they do wrenches. The use of wireless components and communications equipment is on the rise, creating a specific demand for those with knowledge and skills in networking and building automation. The best job opportunities are open to those who have developed troubleshooting skills. Employers are experiencing shortage of qualified applicants who can learn the new technology.

The ongoing focus on energy conservation has prompted the development of new energy-saving HVACR systems. Businesses and home owners want to keep systems operating at peak efficiency. A related issue is pollution reduction. Regulations prohibiting the discharge of CFC and HCFC refrigerants were put in place two decades ago, but the migration to systems that only use new environmentally safe refrigerants is still in progress. At the same time, there has been a continuing concern for improving overall indoor air quality. This growing emphasis on better energy management and pollution control means more HVACR technicians will be needed to retrofit, upgrade, or entirely replace existing climate control systems.

Career Diversity

The industry offers opportunities well beyond technician. One industry expert has counted over 60 different jobs within the field. Careerists may start out as a maintenance technician, but where they end is up to them. They could become a dispatcher, fleet manager, project manager, service manager, product developer, or contractor. They could specialize in distribution, surveying, research and development, energy management, manufacturing, system accessories, or engineering.

There will also be new kinds of jobs in the near future. For example, some of the hot new jobs in the industry include energy auditor, green-technology specialist, and equipment performance-testing specialist.

GETTING STARTED

IT IS WIDELY ACCEPTED THAT THE BEST TIME to job hunt is spring and fall. However, the need for qualified job applicants is so great, any time is a good time to get started.

Finding a job in the HVACR industry can be simple if you are looking in the right places. Here is a list of the best places to check for HVACR openings.

Technical schools

This is the number one place where technicians get started. Individual employers and trade associations know that a vocational school is the best place to look for new recruits. Talk with your counselor and check in the career center to look for new job openings. These jobs are posted often so be sure to check at least once a week. Schools also regularly host job fairs. Participants at job fairs may include employers looking to hire immediately, as well as union representatives offering information about their local apprentice programs.

Unions

You do not need any advance education or training to get into a union-sponsored apprenticeship program. If you are 18 or older and have a high school diploma or equivalent, you qualify. There will be a test, but do not worry – it is only an aptitude test.

An apprenticeship is great because you can earn while you learn. A typical apprenticeship lasts for four years. During that time you will spend some time in the classroom, but more time in the field working under the supervision of an experienced journeyman or master technician.

Trade organizations

Professional associations are excellent resources for those looking to break into the industry. You will find job listings on their websites, but do not stop there. Information about local events will also be posted. You only need to attend one of these events to meet just about everyone in your area who could help you get started. Be prepared with an "elevator speech" and gather as many business cards as you can.

Recruiters

General employment agencies often have job leads for HVACR professionals. There are also agencies and recruiters who specialize in the trades. Start there first before throwing a wider net.

Help wanted

You can find job openings online at general job boards and Craigslist. Be sure and look for job boards that are devoted to the trades, or even limited to the HVACR industry. You can also find help wanted ads in trade publications and your local newspaper.

Social media

Facebook and LinkedIn can be valuable resources. Both sites have groups for people in the HVACR industry. Join the groups and let other members know what you are looking for. You will also find individual companies that may have job openings.

Go direct

Contact potential employers by phone or email. The best companies to look for are HVACR contractors, distributors, and manufacturers.

Someone in the trade

Nothing works better than a referral. Don't know anyone in the HVACR field? No problem. Call local companies and explain that you would like to job shadow one of their technicians for a day. When you shadow, be attentive, helpful, and enthusiastic. At the end of the day, ask for a referral. It can be as simple as that.

And, you are on your way to a great career!

ASSOCIATIONS

■ **Air Conditioning Contractors of America**
http://www.acca.org

■ **American Society of Heating, Refrigerating and Air-Conditioning Engineers (ASHRAE)**
https://www.ashrae.org

■ **Plumbing Heating Cooling Contractors of America**
http://www.phccweb.org

■ **Refrigeration Service Engineers Society**
http://www.rses.org

■ **Sheet Metal and Air Conditioning Contractor National Association**
https://www.smacna.org

■ **Mechanical Contractors Association of America**
http://www.mcaa.org

PERIODICALS

■ **HVAC Insider**
http://www.hvacinsider.com

■ **Air Conditioning, Heating, and Refrigeration News**
http://www.achrnews.com

■ **Contracting Business**
www.contractingbusiness.com

WEBSITES

■ **HVACR Agent**
https://www.hvacagent.com

■ **Air Conditioning, Heating, & Refrigeration Institute**
http://www.ahrinet.org

■ **HVAC Excellence**
http://www.hvacexcellence.org/

■ **HVAC Schools Guide**
http://www.hvacschoolsguide.com

www.ingramcontent.com/pod-product-compliance
Lightning Source LLC
Chambersburg PA
CBHW061236180526
45170CB00003B/1315